The American West

美國西部探險

商務印書館

出版説明

本館一向倡導優質閱讀，近年來連續推出了以“Q”為標識的“Quality English Learning 優質英語學習”系列，其中《讀名著學英語》叢書，更是香港書展入選好書，讀者反響令人鼓舞。推動社會閱讀風氣，推動英語經典閱讀，藉閱讀拓廣世界視野，提高英語水平，已經成為一種潮流。

然良好閱讀習慣的養成非一日之功，大多數初、中級程度的讀者，常視直接閱讀厚重的原著為畏途。如何給年輕的讀者提供切實的指引和幫助，如何既提供優質的學習素材，又提供名師的教學方法，是當下社會關注的重要問題。針對這種情況，本館特別延請香港名校名師，根據多年豐富的教學經驗，精選海外適合初、中級英語程度讀者的優質經典讀物，有系統地出版了這套叢書，名為《Black Cat 優質英語階梯閱讀》。

《Black Cat 優質英語階梯閱讀》體現了香港名校名師堅持經典學習的教學理念，以及多年行之有效的學習方法。既有經過改寫和縮寫的經典名著，又有富創意的現代作品；既有精心設計的聽、説、讀、寫綜合練習，又有豐富的歷史文化知識；既有彩色插圖、繪圖和照片，又有英美專業演員朗讀作品的 CD。適合口味不同的讀者享受閱讀之樂，欣賞經典之美。

《Black Cat 優質英語階梯閱讀》由淺入深，逐階提升，好像參與一個尋寶遊戲，入門並不難，但要真正尋得寶藏，需要投入，更需要堅持。只有置身其中的人，才能體味純正英語的魅力，領略得到真寶的快樂。當英語閱讀成為自己生活的一部分，英語水平的提高自然水到渠成。

商務印書館（香港）有限公司
編輯部

使用説明

1 應該怎樣選書？

按閱讀興趣選書

《Black Cat 優質英語階梯閱讀》精選世界經典作品，也包括富於創意的現代作品；既有膾炙人口的小説、戲劇，又有非小説類的文化知識讀物，品種豐富，內容多樣，適合口味不同的讀者挑選自己感興趣的書，享受閱讀的樂趣。

按英語程度選書

《Black Cat 優質英語階梯閱讀》現設 Level 1 至 Level 6，由淺入深，涵蓋初、中級英語程度。讀物分級採用了國際上通用的劃分標準，主要以詞彙（vocabulary）和結構（structures）劃分。

Level 1 至 Level 3 出現的詞彙較淺顯，相對深的核心詞彙均配上中文解釋，節省讀者查找詞典的時間，以專心理解正文內容。在註釋的幫助下，讀者若能流暢地閱讀正文內容，就不用擔心這一本書程度過深。

Level 1 至 Level 3 出現的動詞時態形式和句子結構比較簡單。動詞時態形式以現在時（present simple）、現在時進行式（present continuous）、過去時（past simple）為主，句子結構大部分是簡單句（simple sentences）。此外，還包括比較級和最高級（comparative and superlative forms）、可數和不可數名詞（countable and uncountable nouns）以及冠詞（articles）等語法知識點。

Level 4 至 Level 6 出現的動詞時態形式，以現在完成時（present perfect）、現在完成時進行式（present perfect continuous）、過去完成時（past perfect continuous）為主，句子結構大部分是複合句（compound sentences）、條件從句（1st and 2nd conditional sentences）等。此外，還包括情態動詞（modal verbs）、被動形式（passive forms）、動名詞（gerunds）、

短語動詞（phrasal verbs）等語法知識點。

根據上述的語法範圍，讀者可按自己實際的英語水平，如詞彙量、語法知識、理解能力、閱讀能力等自主選擇，不再受制於學校年級劃分或學歷高低的約束，完全根據個人需要選擇合適的讀物。

2 怎樣提高閱讀效果？

閱讀的方法主要有兩種：一是泛讀，二是精讀。兩者各有功能，適當地結合使用，相輔相成，有事半功倍之效。

泛讀，指閱讀大量適合自己程度（可稍淺，但不能過深）、不同內容、風格、體裁的讀物，但求明白內容大意，不用花費太多時間鑽研細節，主要作用是多接觸英語，減輕對它的生疏感，鞏固以前所學過的英語，讓腦子在潛意識中吸收詞彙用法、語法結構等。

精讀，指小心認真地閱讀內容精彩、組織有條理、遣詞造句又正確的作品，着重點在於理解“準確”及“深入”，欣賞其精彩獨到之處。精讀時，可充分利用書中精心設計的練習，學習掌握有用的英語詞彙和語法知識。精讀後，可再花十分鐘朗讀其中一小段有趣的文字，邊唸邊細心領會文字的結構和意思。

《Black Cat 優質英語階梯閱讀》中的作品均值得精讀，如時間有限，不妨嘗試每兩個星期泛讀一本，輔以每星期挑選書中一章精彩的文字精讀。要學好英語，持之以恆地泛讀和精讀英文是最有效的方法。

3 本系列的練習與測試有何功能？

《Black Cat 優質英語階梯閱讀》特別注重練習的設計，為讀者考慮周到，切合實用需求，學習功能強。每章後均配有訓練聽、説、讀、寫四項技能的練習，分量、難度恰到好處。

聽力練習分兩類，一是重聽故事回答問題，二是聆聽主角對話、書信朗讀、或模擬記者訪問後寫出答案，旨在以生活化的練習形式逐步提高聽力。每本書均配有CD提供作品朗讀，朗讀者都是專業演員，英國作品由英國演員錄音，美國作品由美國演員錄音，務求增加聆聽的真實感和感染力。多聆聽英式和美式英語兩種發音，可讓讀者熟悉二者的差異，逐漸培養分辨英美發音的能力，提高聆聽理解的準確度。此外，模仿錄音朗讀故事或模仿主人翁在戲劇中的對白，都是訓練口語能力的好方法。

閱讀理解練習形式多樣化，有縱橫字謎、配對、填空、字句重組等等，注重訓練讀者的理解、推敲和聯想等多種閱讀技能。

寫作練習尤具新意，教讀者使用網式圖示（spidergrams）記錄重點，採用問答、書信、電報、記者採訪等多樣化形式，鼓勵讀者動手寫作。

書後更設有升級測試（Exit Test）及答案，供讀者檢查學習效果。充分利用書中的練習和測試，可全面提升聽、説、讀、寫四項技能。

4 本系列還能提供甚麼幫助？

《Black Cat 優質英語階梯閱讀》提倡豐富多元的現代閱讀，巧用書中提供的資訊，有助於提升英語理解力，擴闊視野。

每本書都設有專章介紹相關的歷史文化知識，經典名著更有作者生平、社會背景等資訊。書內富有表現力的彩色插圖、繪圖和照片，使閱讀充滿趣味，部分加上如何解讀古典名畫的指導，增長見識。有的書還提供一些與主題相關的網址，比如關於不同國家的節慶源流的網址，讓讀者多利用網上資源增進知識。

CONTENTS

The text is recorded in full. 故事錄音

These symbols indicate the beginning and end of the extracts linked to the listening activities. 聽力練習開始和結束的標記

A NATION IS BORN

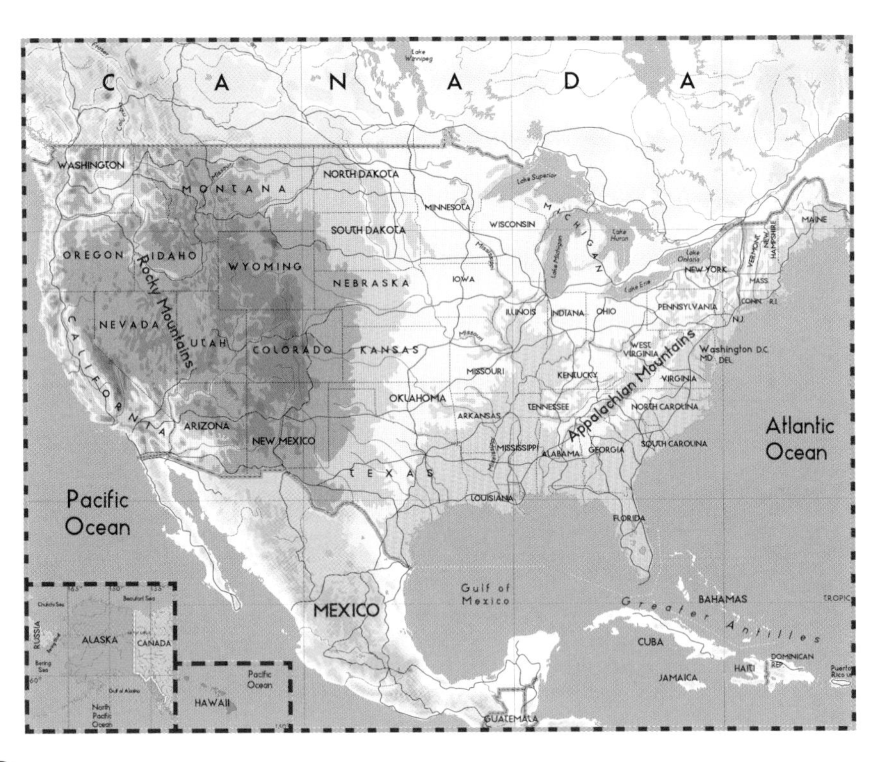

Population: [1] about 250 million people

States: 50

Important Rivers: Mississippi and Missouri

Important Mountains: Rocky and Appalachian

Important Lakes: Superior, Michigan, Huron, Erie, Ontario

This is a map of the United States today. It is a big country with 50 states. [2]

1. **population** : 人口。
2. **states** : 州。

The first European explorers[1] came to America about 400 years ago. At that time America was an enormous[2] wilderness[3] and Indians lived there.

America became independent[4] from Britain in 1776. It became a nation and grew very quickly.

Enthusiastic[5] people wanted to explore and live in the West. The story of the American West is exciting!

KET

Are these sentences "Right" (A) or "Wrong" (B)? If there is not enough information to answer "Right" (A) or "Wrong" (B), choose "Doesn't say" (C).

0. The United States of America has 50 states.

(**A**) Right **B** Wrong **C** Doesn't say

1. The Mississippi and Appalachian are two important rivers.

A Right **B** Wrong **C** Doesn't say

2. America has four important lakes.

A Right **B** Wrong **C** Doesn't say

3. About 250 million people live in the United States.

A Right **B** Wrong **C** Doesn't say

4. The Rocky Mountains are very tall.

A Right **B** Wrong **C** Doesn't say

5. The first European explorers came to America about 400 years ago.

A Right **B** Wrong **C** Doesn't say

6. The Indians explored the West.

A Right **B** Wrong **C** Doesn't say

1. **explorers** : 探險者。
2. **enormous** : 廣大的。
3. **wilderness** : 荒野。
4. **independent** : 獨立的。
5. **enthusiastic** : 熱情的。

BEFORE YOU READ

Do you know these words?

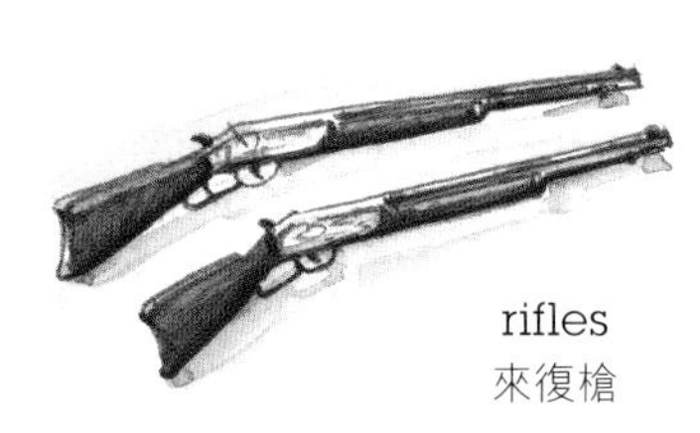

rifles
來復槍

cattle
牛

stagecoach
公共馬車

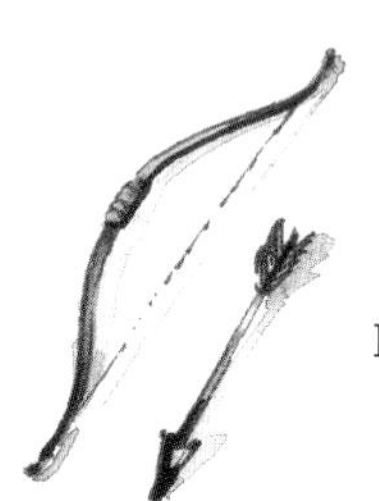

bow and arrow
弓與箭

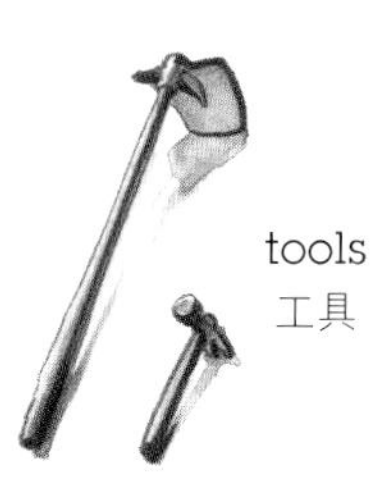

tools
工具

sheep
羊

tepee
（北美印第安人用樹皮、獸皮等製成的）圓錐形帳篷

covered wagon
有篷蓋的手推車

KET

2 Listen to Chapter One and answer the following questions. Tick (✓) A, B or C.

0. How many Indians lived in America when the first explorers arrived?
   - A ☐ 40,000.
   - B ☐ 30 million.
   - C ☑ 20 million.

1. What did every Indian tribe have?
   - A ☐ A buffalo.
   - B ☐ A chief.
   - C ☐ Tobacco plants.

2. Who hunted the buffalo?
   - A ☐ The Indians of the Plains.
   - B ☐ All the Indians.
   - C ☐ The explorers.

3. Who were expert hunters?
   - A ☐ The Indians of the West.
   - B ☐ The Indians of the forest.
   - C ☐ The Indians of the Plains.

4. Who loved and respected nature?
   - A ☐ The explorers.
   - B ☐ The Indians.
   - C ☐ The Indian chiefs.

5. Where did the settlers want to go?
   - A ☐ To the West.
   - B ☐ To Asia.
   - C ☐ To South America.

6. Why were the Indians very angry?
   - A ☐ Because they had no food.
   - B ☐ Because they had no rifles.
   - C ☐ Because they lost their land and their way of life.

7. What are reservations?
   - A ☐ Indian lands in the West.
   - B ☐ Lands with lots of buffalo.
   - C ☐ Government land for the Indians to live on.

CHAPTER ONE

The First Americans: The Indians

The American Indians came from Asia about 40,000 years ago! They crossed the Bering Strait.[1] Then they went down to North and South America. (See map on page 13) About 20 million Indians lived in America when the first explorers arrived. These Indians lived in many different tribes.[2] Every tribe had a chief.[3]

The Indians and the explorers were friends. The Indians taught the explorers how to grow corn,[4] potatoes and tobacco.[5] They taught them how to travel by canoe[6] in the wilderness.

1. **strait** : 海峽。
2. **tribes** : 部落。
3. **chief** : 酋長。
4. **corn** : 
5. **tobacco** : 煙草。
6. **canoe** : 獨木舟。

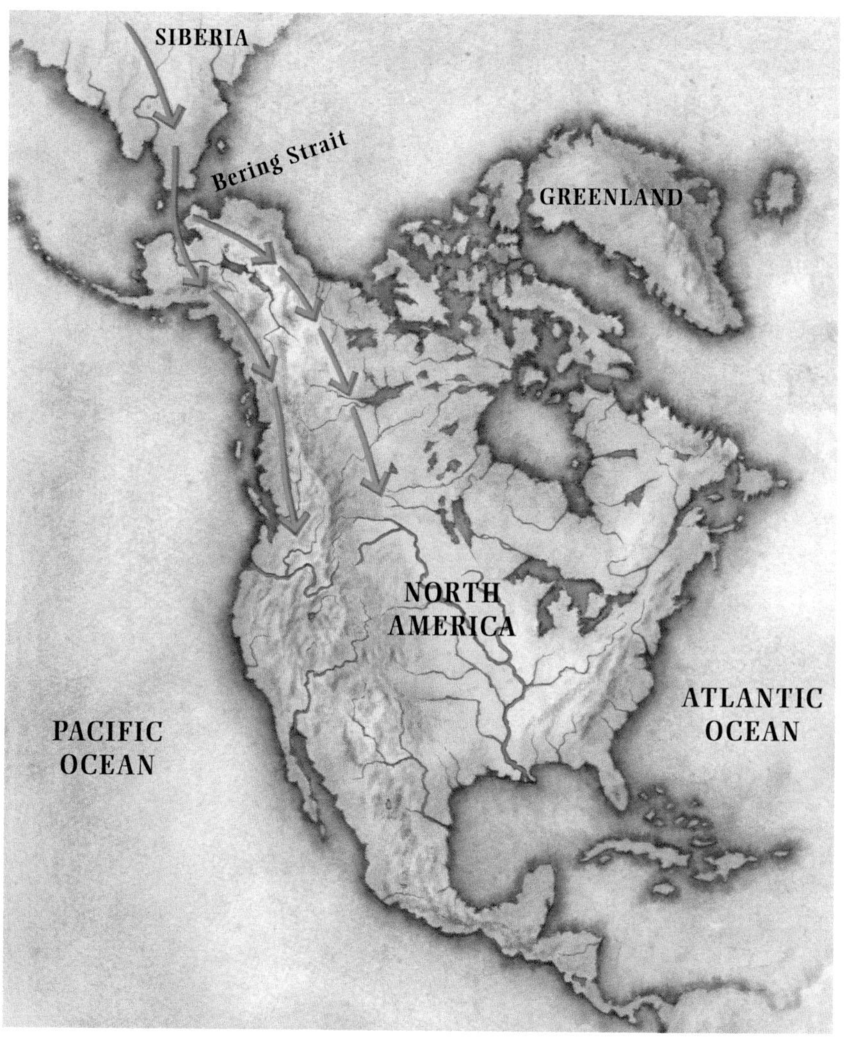

The arrows on the map show how the American Indians came from Siberia (Asia) to North America. They crossed the Bering Strait and then traveled to different parts of North America. With time they settled on the North American continent.

The explorers brought the Indians rifles, tools, cattle and horses. The explorers also brought illnesses.[1] Many Indians died from these illnesses.

The Indians of the Plains[2] were expert hunters.[3] They rode fast horses. They hunted the buffalo[4] with bows and arrows or rifles.

These Indians followed the buffalo across the Plains. They used the buffalo for food, clothes and tepees.

The other American Indians were farmers. They farmed, fished and hunted small animals.

The Great Buffalo Hunt by William Robinson Leigh, 1947.
(The Rockwell Museum, New York)

1. **illnesses**：（單數形式為 illness）疾病。
2. **plains**：平原。
3. **expert hunters**：技藝高超的獵手。
4. **buffalo**：水牛。

These native [1] Americans were happy people. They were strong, healthy [2] and courageous. [3] They loved and respected nature.

After many years, more new settlers and explorers went to America. They wanted to go to the West. They wanted to live there. They took the land from the Indians. The Indians were very angry and the Indian Wars started. Many people were killed.

The Indians lost their land and their way of life. They went to live on reservations. [4]

The Captive Charger by Charles Wimar, 1854.
(The Saint Louis Museum of Art, St. Louis, Missouri)

1. **native**：本地的。
2. **healthy**：身體健康的。
3. **courageous**：勇敢的。
4. **reservations**：北美印第安人的居留地。

UNDERSTANDING THE TEXT

KET

Are these sentences "Right" (A) or "Wrong" (B)? If there is not enough information to answer "Right" (A) or "Wrong" (B), choose "Doesn't say" (C).

0. The first Indians came from Asia about 40,000 years ago.
(**A**) Right **B** Wrong **C** Doesn't say

1. They crossed the Bering Strait during the summer.
A Right **B** Wrong **C** Doesn't say

2. The Indians taught the explorers to cure illnesses with herbs.
A Right **B** Wrong **C** Doesn't say

3. The Indians lived together in one big tribe.
A Right **B** Wrong **C** Doesn't say

4. The explorers brought the Indians rifles, cattle and horses.
A Right **B** Wrong **C** Doesn't say

5. The Indians did not like the rifles.
A Right **B** Wrong **C** Doesn't say

6. The Indians hunted the buffalo on the Plains.
A Right **B** Wrong **C** Doesn't say

7. They used the buffalo to make bows and arrows.
A Right **B** Wrong **C** Doesn't say

8. The settlers took the land from the Indians.
A Right **B** Wrong **C** Doesn't say

9. Many people were killed during the Indian Wars.
A Right **B** Wrong **C** Doesn't say

Complete the table with the correct verb forms.

INFINITIVE（不定式）	PAST SIMPLE（過去式）
lose	
	took
want	
love	
	used
follow	
ride	
	died
bring	
	taught
	was/were
have	
	lived
	went
come	

Use some of the verbs above to fill in the gaps.

a. About 20 million Indians .... in America.

b. Every tribe .... a chief.

c. The Indians .... the explorers how to grow corn.

d. The explorers .... the Indians cattle and horses.

e. The Indians of the Plains .... fast horses.

f. The new settlers and explorers .... to go to the West.

g. The Indians .... their land.

Odd one out!
Circle the word that doesn't belong to the group.

a. Asia America reservations Europe

b. Europeans Indians Americans hunters

c. food home tepee castle

d. sheep canoe buffalo horses

e. courageous strong healthy chief

f. tools corn potatoes apples

5 **Now use the odd words to fill the gaps.**

a. The explorers brought the Indians ....

b. Each tribe had a ....

c. The Indians were expert ....

d. They used the buffalo for ....

e. They traveled by ....

f. The Indians went to live on ....

T: GRADE 3

Topic – Friends
"The Indians and the explorers were friends."
Use a photo of your best friend and answer the following questions:

a. What's the name of your best friend?

b. Describe him/her.

c. What do you do together at the weekend?

Daniel Boone

Before 1803 most of America was a wilderness. There were no maps and no roads.

Daniel Boone was a famous explorer and pioneer.[1] In 1769 he traveled across the Appalachian Mountains to Kentucky. He opened the road to the West.

Daniel Boone and his family built a cabin[2] and a fort[3] in Kentucky. Boone had many adventures[4] with the Indians. The Indians captured[5] and killed two of his sons. In 1778 Boone was captured by the Shawnee Indians. He was their prisoner but these Indians respected him. He was a strong fighter and a courageous man.

Boone helped the pioneers. His fort protected settlers during Indian attacks. Boone was an American free spirit. He loved freedom and adventure. In 1799 he went to Missouri. He said to

1. **pioneer** : 拓荒者。
2. **cabin** : 小木屋。
3. **fort** : 城堡。
4. **adventures** : 冒險活動。
5. **captured** : 俘虜。

his friends, "There are too many people here. I'm going to the West!" America grew very quickly.

Johnny Appleseed was another American pioneer. In the 1800s he planted apple trees on the frontier. He helped the settlers in Ohio and Indiana.

Look at this table:

THE RAPID GROWTH[1] OF AMERICA

YEAR	POPULATION
1781*	3 million
1819	9.3 million
1849	22 million
1869	39 million

*at the end of the American Revolution

Daniel Boone Escorting[2] Settlers Through the Cumberland Gap[3]
by George Caleb Bingham, 1852.
(Washington University Gallery of Art, St. Louis, Missouri)

1. **growth** : 興起。 2. **escorting** : 護送。 3. **gap** : 峽谷。

BLACK CAT ENGLISH CLUB

Membership Application Form

BLACK CAT ENGLISH CLUB is for those who love English reading and seek for better English to share and learn with fun together.

Benefits offered:
- *Membership Card*
- *Book discount coupon*
- *English learning e-forum*
- *English learning activities*
- *Black Cat English Reward Scheme*
- *Surprise gift and more...*

Simply fill out the application form below and fax it back to 2565 1113 or send it back to the address at the back.

Join Now! It's FREE exclusively for readers who have purchased *Black Cat English Readers* **!**

(Please fill out the form with **BLOCK LETTERS**.)

The title of Black Cat English Reader/book set that you have purchased: ____________

English Name: ____________ (Surname) ____________ (Given Name)

Chinese Name: ____________

Address: ____________

Tel: ____________ Fax: ____________

Email: ____________

(Login password for e-forum will be sent to this email address.)

Sex: ❑ Male ❑ Female

Education Background: ❑ Primary 1-3 ❑ Primary 4-6 ❑ Junior Secondary Education (F1-3) ❑ Senior Secondary Education (F4-5) ❑ Matriculation ❑ College ❑ University or above

Age: ❑ 6 - 9 ❑ 10 - 12 ❑ 13 - 15 ❑ 16 - 18 ❑ 19 - 24 ❑ 25 - 34 ❑ 35 - 44 ❑ 45 - 54 ❑ 55 or above

Occupation: ❑ Student ❑ Teacher ❑ White Collar ❑ Blue Collar ❑ Professional ❑ Manager ❑ Business Owner ❑ Housewife ❑ Others (please specify: ____________)

As a member, what would you like **BLACK CAT ENGLISH CLUB** to offer:

❑ Member gathering/ party ❑ English class with native teacher ❑ English competition
❑ Newsletter ❑ Online sharing ❑ Book fair
❑ Book discount ❑ Others (please specify: ____________)

Other suggestions to **BLACK CAT ENGLISH CLUB**: ____________

Please sign here: ____________ (Date: ____________)

Visit us at Quality English Learning Online http://publish.commercialpress.com.hk/qel

Stamp
Here

BLACK CAT ENGLISH CLUB
The Commercial Press (Hong Kong) Ltd.
9/F, Eastern Central Plaza,
3 Yiu Hing Road, Shau Kei Wan,
Hong Kong

UNDERSTANDING THE TEXT

1 Read the sentences about Chapter Two. Choose the best word (A, B or C) for each space.

0. There were no roads in America .... 1803.

A after (**B**) before **C** during

1. Daniel Boone was a famous ....

A spirit **B** prisoner **C** explorer

2. Boone opened the .... to the West.

A street **B** map **C** road

3. Boone .... a fort in Kentucky.

A built **B** did **C** made

4. He was a .... of the Shawnee Indians.

A fighter **B** teacher **C** prisoner

5. Johnny Appleseed .... apple trees on the frontier.

A planted **B** put **C** had

2 LOOKING AT PICTURES
Look at the painting on page 20 and answer these questions.

a. Who are the people in the picture?

....

b. Where are they going?

....

c. What season is it?

....

d. What's the weather like?

....

KET

Complete the five conversations. Choose the correct answers (A, B or C).

0. What day is it today?

A ☐ It's the tenth.
B ☐ It's May.
C ☑ It's Tuesday.

1. How many horses have you got?

A ☐ Eight.
B ☐ Any.
C ☐ Nothing.

2. Are you going to the fort tomorrow?

A ☐ No, I'm not.
B ☐ No, I don't.
C ☐ Yes, I do.

3. John is a free spirit.

A ☐ I haven't.
B ☐ I am too.
C ☐ I don't.

4. Where is the nearest house?

A ☐ I don't know.
B ☐ Two hours.
C ☐ Yes, there is.

5. What time is breakfast?

A ☐ After one hour.
B ☐ In the morning.
C ☐ At eight o'clock.

Dictation

Listen to the dictation carefully. Then listen to it again and fill in the gaps with the words you hear.

Daniel Boone was a famous pioneer. He opened the .... to the West. He built a cabin and a .... in Kentucky.
Boone always helped the pioneers. He .... freedom and adventure. He was an American .... spirit. In 1799 he .... to Missouri. He said to his friends, "There are too .... people here. I'm going to the .... !"

Word search

Look at the places on the map on page 8. Then look below and circle:

a. two important American rivers
b. two important American mountains
c. one important American lake

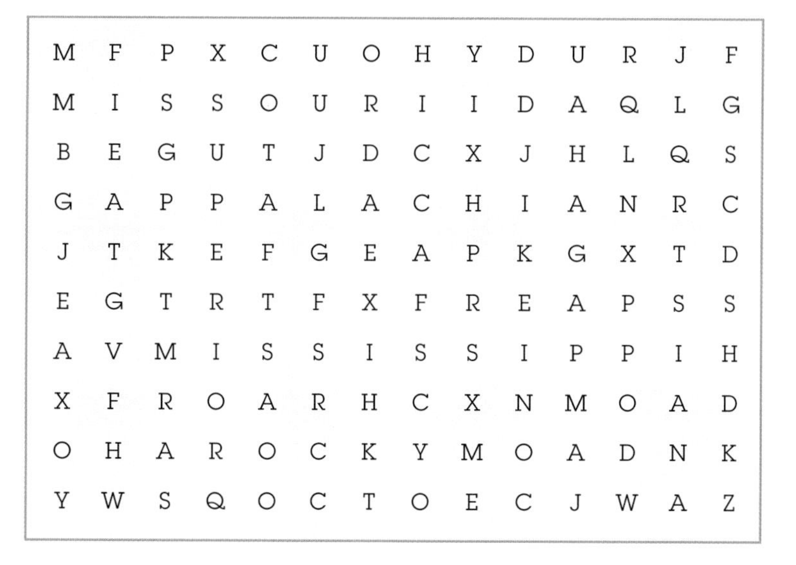

M	F	P	X	C	U	O	H	Y	D	U	R	J	F
M	I	S	S	O	U	R	I	I	D	A	Q	L	G
B	E	G	U	T	J	D	C	X	J	H	L	Q	S
G	A	P	P	A	L	A	C	H	I	A	N	R	C
J	T	K	E	F	G	E	A	P	K	G	X	T	D
E	G	T	R	T	F	X	F	R	E	A	P	S	S
A	V	M	I	S	S	I	S	S	I	P	P	I	H
X	F	R	O	A	R	H	C	X	N	M	O	A	D
O	H	A	R	O	C	K	Y	M	O	A	D	N	K
Y	W	S	Q	O	C	T	O	E	C	J	W	A	Z

The Lewis and Clark Expedition[1]

In 1803 the United States President,[2] Thomas Jefferson, bought the Louisiana territory[3] from France. He paid $15 million for this immense[4] piece of land.

President Jefferson liked science. It was very interesting. He wanted to learn about the plants and animals of the new territory.

In May 1804 President Jefferson asked Meriwether Lewis and William Clark to explore the Louisiana Purchase.[5] Lewis and Clark were the first white men to see this land.

1. **expedition** : 探險。
2. **president** : 總統。
3. **territory** : 地區。
4. **immense** : 巨大的。
5. **purchase** : 購買的土地。

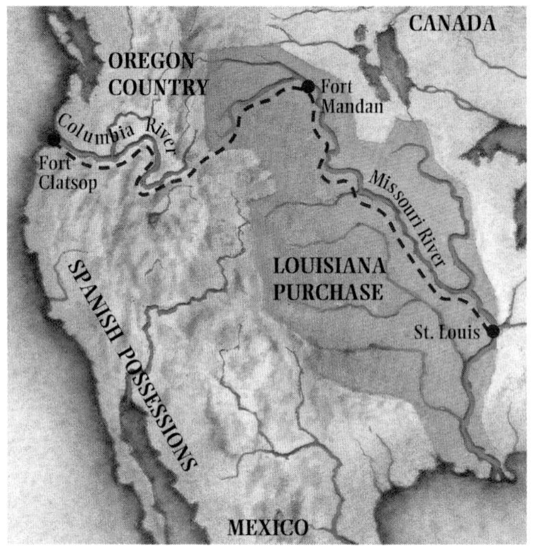

This important expedition began in St. Louis, Missouri. About 50 men traveled up the Missouri River. They explored the territory and made maps.

After six months they stopped and built Fort Mandan. They spent the long winter there.

In the spring of 1805 the expedition continued. An Indian woman called Sacagawea went on the expedition with the explorers. She was a friendly Shoshone Indian.

Sacagawea spoke English and other Indian languages. She was very helpful. Sacagawea's brother was a Shoshone chief. A Shoshone guide helped the explorers cross the high mountains. The journey was difficult and dangerous.[1] Grizzly bears[2] and other animals attacked the explorers.

The explorers built canoes. They traveled down the Columbia River to the Pacific Ocean! They arrived at the Pacific Ocean in November 1805. They built a fort there.

The expedition returned to Missouri in September 1806. The explorers traveled 8,000 miles (12,800 kilometers) from May 1804 to September 1806.

1. **dangerous**：危險的。

2. **grizzly bears**：灰熊。

The Lewis and Clark expedition was very successful. Lewis and Clark made many maps. They brought back a lot of information about the West.

In the 1840s John Fremont and Kit Carson explored the land between the Rocky Mountains and the Pacific Ocean. They visited California and Oregon. Fremont made the first scientific map of the West.

Another important pioneer was Davy Crockett. He fought for the independence [1] of Texas.

This was the beginning of the great movement [2] to the West.

The expedition spent the winter at Fort Mandan.

1. **independence**：獨立。
2. **movement**：遷移。

UNDERSTANDING THE TEXT

Choose the correct words and circle them.

a. In 1803 President Thomas Jefferson *bought* / *sold* the Louisiana territory for $15 million.

b. In 1804 Lewis and Clark went to *live in* / *explore* the Louisiana Purchase.

c. About *fifteen* / *fifty* men traveled *up* / *down* the Missouri River.

d. They spent the *winter* / *year* in Fort Mandan.

e. Sacagawea was a friendly Shoshone Indian *chief* / *woman*.

f. In 1805 the Lewis and Clark expedition arrived at the Pacific *River* / *Ocean*.

g. John Fremont and Kit Carson were two important *explorers* / *Indians*.

Match the adjectives with the correct situations.

e.g. When you see your friend you are happy.

Situations	Adjectives
a. When you run a lot	**1.** you are thirsty.
b. When the summer holiday ends	**2.** you are frightened.
c. When you go to the library	**3.** you are hungry.
d. When you want to drink	**4.** you are sad.
e. When you want to eat	**5.** you are tired.
f. When you see a grizzly bear	**6.** you are quiet.

3 **Descriptions**
Look at these characters. Use the words in the box below to describe them.

courageous friendly helpful
explorers made maps built a fort Indian
spoke many Indian languages

M. Lewis and W. Clark	Sacagawea
....	....
....	....
....	....
....	....

Let's go West!

After the Lewis and Clark expedition settlers, gold prospectors [1] and trappers [2] went to the West. The American frontier [3] was immense.

Settlers wanted to build homes and farms in the West. They wanted a better life. Gold prospectors wanted to find gold and become rich. Trappers wanted to hunt wild animals. This was an exciting period.

In 1841 thousands of pioneers began their long journey. They started in Independence, Missouri. From there they took the

1. **gold prospectors** : 淘金者。
2. **trappers** : （為獲得野獸皮毛而）設陷阱捕獸者。
3. **frontier** : 邊疆。

The long wagon train. The Jerkline by Charles M. Russell, 1912.
(C. M. Russell Museum, Montana)

Oregon Trail [1] or the Santa Fe Trail. They traveled for four to six months.

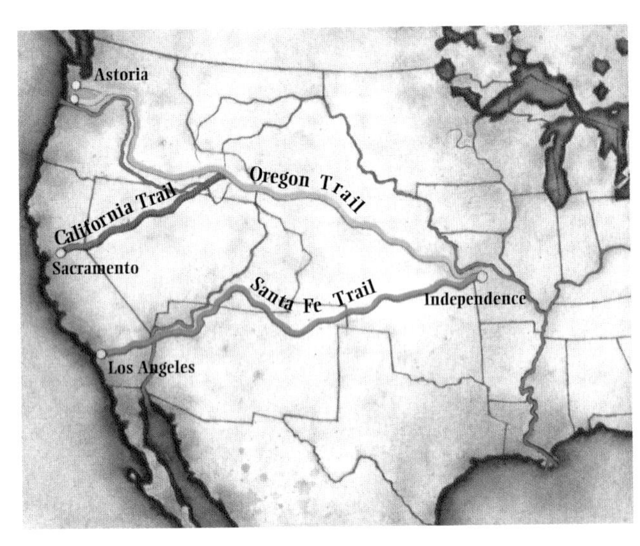

Most journeys began in the spring. The pioneers wanted to arrive before winter. The pioneers traveled in covered wagons pulled by mules [2] or oxen. [3] They put food, clothes, furniture [4] and other things in the covered wagon. A lot of families took cattle and sheep.

Many covered wagons traveled together. This was called a wagon train. All the wagon trains had a leader called a captain. [5] They also had a scout. [6] The scout knew the trail well. He walked in front of the others to look for Indians or other dangers.

The journey was long, difficult and dangerous. It was difficult to cross big rivers and tall mountains.

The weather was another problem. In the summer it was very hot. There was little water to drink. There was little grass for the animals to eat. Rain and snow were also a problem.

About 10,000 people died on the Oregon Trail between 1835 and 1845. The Indians killed only 400 people!

1. **trail**：路線。
2. **mules**：
3. **oxen**：（單數形式為 ox）
4. **furniture**：家具。
5. **captain**：車長。
6. **scout**：偵察兵。

A TYPICAL DAY ON THE TRAIL:

5 a.m.: The pioneers got up and had breakfast. Then they began to travel.

Midday: The pioneers and the animals rested. They drank water and ate.

2 p.m.: The wagon trains began to travel again.

Sunset:[1] At sunset the scout chose a campsite.[2] The wagon train made a big circle to protect everyone from wild animals. In the center of the circle there were campfires.[3]

The pioneers sat around the campfires to eat and talk. They went to bed early because they were tired. A day on the trail wasn't easy!

A pioneer and his cattle.
The Stampede[4] by Frederic Remington, 1910.
(Remington Art Museum, New York)

1. **sunset** : 
2. **campsite** : 野營地。
3. **campfires** : 
4. **stampede** : （美國西北部）牛仔的競技表演。

UNDERSTANDING THE TEXT

KET

Read the paragraph and choose the best word (A, B or C) for each space.

After the Lewis and Clark expedition [0]...C..... people went to the West. The long journey [1].......... in Independence, Missouri. The pioneers traveled [2].......... covered wagons. They [3].......... the Oregon Trail or the Santa Fe Trail. [4].......... wagon train had a captain and a scout.

[5].......... were many problems [6].......... the journey. The weather [7].......... a big problem.

[8].......... 1835 and 1845 ten thousand people died on the Oregon Trail. A typical day on the trail wasn't [9].......... .

	A	B	C
0.	**A** lot	**B** much	**C** many
1.	**A** began	**B** beginning	**C** begin
2.	**A** by	**B** in	**C** at
3.	**A** went	**B** got	**C** took
4.	**A** Every	**B** Any	**C** An
5.	**A** There	**B** Their	**C** They're
6.	**A** of	**B** during	**C** about
7.	**A** was	**B** is	**C** were
8.	**A** At	**B** Among	**C** Between
9.	**A** well	**B** easy	**C** good

T: GRADE 3

Topic – Weather
"The weather was a big problem."
Describe the weather in your country by answering the following questions.

a. What's the weather like in the summer and winter in your country?
b. Does it rain a lot?
c. Is the weather ever a big problem in your country?
d. What was the weather like yesterday?
e. What type of weather do you prefer?

KET

Complete the conversation. Jack meets Sue on the wagon train. What does Jack say to Sue? Put the correct letters A-H in the spaces.

Jack: Hello! I'm from Boston.

Sue: 0 ..B..

Jack: Where are you going?

Sue: 1 .....

Jack: Do you like traveling on the Oregon Trail?

Sue: 2 .....

Jack: Me too. But do you like this heat?

Sue: 3 .....

Jack: Is that your dog?

Sue: 4 .....

Jack: Can you come to see my covered wagon?

Sue: 5 .....

A Yes, it is.

B I am too.

C No, I'm not.

D Yes, I can.

E To California.

F You're welcome.

G No, it's too hot for me.

H Yes, I do.

ODD ONE OUT!
Circle the word that doesn't belong to the group.

a. settler scientist scout prospector

b. rich important famous poor

c. trail trip journey voyage

d. table chairs wagon furniture

e. difficult danger fun problem

Choose one odd word and make a sentence with it.

....

5 Read the descriptions of the people in the wagon train. What is the word for each one? The first letter is already there. There is one space for each other letter in the word.

0. I look for gold. p r o s p e c t o r
1. I want to build my home in the West. s _ _ _ _ _ _
2. I hunt animals and sell their skins. t _ _ _ _ _ _
3. I'm the leader of a wagon train. c _ _ _ _ _ _
4. I look for Indians and other dangers. s _ _ _ _
5. I pull the big covered wagon. m _ _ _

6 LOOKING AT PICTURES

Look at the painting on pages 30 and 31.

a. What time of day is it?

....

b. What's the weather like?

....

c. What can you see?

....

d. What is the captain of the wagon train doing?

....

e. Do you like this painting?

....

Why or why not?

....

BEFORE YOU READ

KET

8 **1** **Listen to Chapter Five and answer the following questions. Tick (✓) A, B or C.**

0. The settlers began to work
- **A** ☐ after the first year.
- **B** ☑ when they arrived at their destination.
- **C** ☐ when the weather was good.

1. Where did the settlers get their furniture?
- **A** ☐ At Fort Laramie.
- **B** ☐ At Fort Bridger.
- **C** ☐ They made it.

2. The settlers built homes called
- **A** ☐ log houses.
- **B** ☐ log cabins.
- **C** ☐ pioneer cabins.

3. Men, women and children planted
- **A** ☐ crops.
- **B** ☐ flowers.
- **C** ☐ apple trees.

4. The settlers ate a lot of
- **A** ☐ wheat.
- **B** ☐ meat.
- **C** ☐ corn.

5. On the frontier there were no
- **A** ☐ doctors.
- **B** ☐ hospitals.
- **C** ☐ spinning wheels.

6. The "circuit rider" was
- **A** ☐ a scout.
- **B** ☐ an Indian chief.
- **C** ☐ a religious man.

7. Most settlers were
- **A** ☐ honest people.
- **B** ☐ outlaws.
- **C** ☐ women.

Life on the Frontier

When the settlers arrived at their destination [1] they began to work. They began to build a home and plant crops. The home was usually made of logs. [2] It was called a log cabin. It was a small home with one or two rooms. The settlers made the furniture. They used logs to make tables, chairs and beds.

Men, women and children planted crops. Corn and wheat [3] were important crops. The settlers ate a lot of corn. The men hunted buffalo and other animals. Men, women and children worked all day. There was little free time.

1. **destination** : 目的地。
2. **logs** : 
3. **wheat** : 

Frontier people were true American free spirits.
(U.S. Library of Congress)

Abilene, a frontier town in 1879.
(Kansas State Historical Society, Topeka, Kansas)

There were no shops on the frontier. The settlers bought some things from Fort Laramie and Fort Bridger. Some women brought a spinning wheel [1] to the frontier. These women made yarn. [2] They used yarn to make clothes.

Illness was a big danger on the frontier. There were no hospitals, few doctors and very little medicine! Many people died at a young age.

Most settlers lived far from churches. When they wanted to get married, they waited for the "circuit rider". [3] The "circuit rider" was a religious man. He traveled on the frontier. Settlers often waited months for the "circuit rider".

END

Who were the frontier people? The frontier people came from the East, the South and the Indiana Territory. They also came from Europe, Scandinavia and China! Most settlers were honest people but some were outlaws. [4] Some of them wanted excitement [5] and adventure.

Frontier people were strong and courageous. They loved the freedom and adventure of the frontier. They were true American free spirits.

Villages and towns began to grow on the frontier. Some became important cities.

1. **spinning wheel** : 
2. **yarn** :
3. **circuit rider** : 流動的神職人員。
4. **outlaws** : （此處指）逃犯。
5. **excitement** : 刺激。

Pioneers used logs to build homes.

UNDERSTANDING THE TEXT

KET

Are these sentences "Right" (A) or "Wrong" (B)? If there is not enough information to answer "Right" (A) or "Wrong" (B), choose "Doesn't say" (C).

0. When the settlers arrived at their destination they began to build a home.

(**A**) Right **B** Wrong **C** Doesn't say

1. The home was called a fort.

A Right **B** Wrong **C** Doesn't say

2. Corn and bread were important crops.

A Right **B** Wrong **C** Doesn't say

3. The settlers bought some things from Fort Laramie and Fort Bridger.

A Right **B** Wrong **C** Doesn't say

4. Abilene was the first frontier town.

A Right **B** Wrong **C** Doesn't say

5. Not many people died at a young age.

A Right **B** Wrong **C** Doesn't say

6. Most settlers lived far from churches.

A Right **B** Wrong **C** Doesn't say

Frontier people loved freedom and adventure.

Look at this sentence from Chapter 5:

The home was *usually* made of logs.

Usually **is an adverb of frequency（表示活動頻率的副詞）. We use adverbs of frequency to say how often something happens.**

Look at the frequency adverbs below. Then write what you do in your free time by putting the phrases under the correct adverbs. You will find the phrases in the box.

listen to music　watch TV　read a book　meet friends
visit my grandparents　eat a pizza　play a musical instrument
sleep　phone a friend　go to the cinema　play football
listen to the radio　go to the disco　go shopping

always	usually
....	....
....	....
....	....
....	....

sometimes	never
....	....
....	....
....	....
....	....

T: GRADE 3

Topic – Free time
"Men, women and children worked all day. There was little free time."
Think about what you do in your free time.

a. What do you usually do after school?

b. What do you do on Sundays?

c. What do you do in the summer holidays?

d. Would you like to have more free time?

KET

4 **You are a young settler. You and your family traveled on the Oregon Trail. Now you are in the Oregon Territory. You want to write a letter to your best friend in Kentucky. Write one word for each space.**

Dear Tom,

I traveled 1 .... the Oregon Trail 2 ....
five months. My wagon train was attacked 3 .... the Indians. I was very frightened. It 4 .... very hot on the trail. 5 .... was little water to drink.

6 .... father built a home. We work 7 .... day. Yesterday we went 8 .... the fort to buy some things. We traveled for three hours 9 .... our covered wagon. There 10 .... wild animals near the fort.

Your friend,

....

5 **LOOKING AT PICTURES**

Look at the picture on page 41 and answer the following questions.

a. What are the men doing?

....

b. What are the women doing?

....

c. How many men, women and children are there?

Men .... Women .... Children ....

6 Listen to the first five paragraphs of Chapter 5 and put the pictures in the order that they are mentioned.
Write 1, 2, 3 etc. in the correct box.

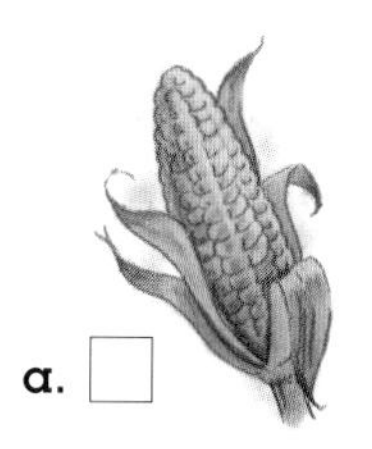

a. ☐

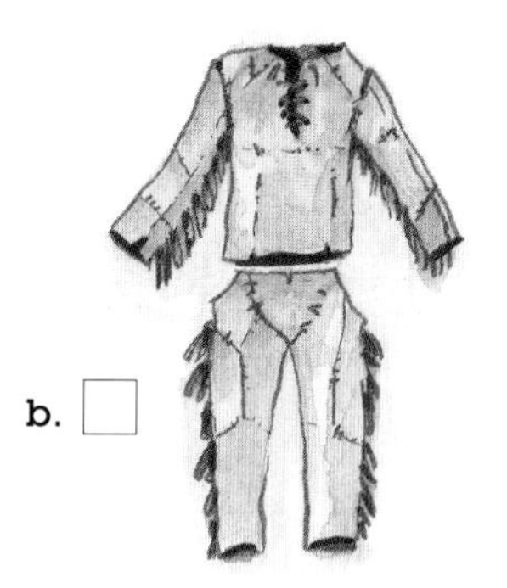

b. ☐

c. ☐

d. ☐

e. ☐

f. ☐

OUTLAWS OF THE WEST

It was very difficult to keep law and order [1] in the West. Most settlers were honest people, but there were a lot of outlaws. Outlaws robbed banks, trains and stagecoaches. Honest people were afraid to travel. There were horse thieves [2] and cattle thieves, but there were few sheriffs! [3]

Who were the sheriffs of the West? The most famous were Wyatt Earp, Pat Garrett, Wild Bill Hickock and Bat Masterson. Wyatt Earp and his brother Virgil were Western heroes. In October 1881 they killed three outlaws at the O.K. Corral in Arizona. The gunfight at the O.K. Corral became a legend. There are several films about this gunfight.

The O.K. Corral in Tombstone, Arizona.

Who were the outlaws of the West? There were a lot of outlaws in the West. The most famous were Billy the Kid and Jesse James.

Billy the Kid was a killer and a cattle thief. He worked on a cattle ranch [4] in New Mexico.

1. **law and order** : 法律與秩序。
2. **thieves** : （單數形式為 thief）賊。
3. **sheriffs** : （美國）縣治安官。
4. **ranch** : （美國）大牧場。

The $5,000 reward for Billy the Kid.

There was a $5,000 reward [1] for Billy the Kid. In 1880 Sheriff Pat Garrett captured him and put him in prison. Billy escaped from prison. He killed two of the sheriff's men. Sheriff Pat Garrett followed him to Fort Sumner, New Mexico.

On July 14 1881, he killed Billy the Kid during a gunfight. Billy was only 21 years old!

Jesse James and his brother Frank were bank and train robbers. They terrorized [2] the West for many years.

In 1881 there was a $10,000 reward for the arrest of Jesse or Frank James. On April 3 1882, an outlaw called Bob Ford killed Jesse in Missouri.

In the West horse thieves, cattle thieves and killers were usually hanged [3] when they were captured. The West was a violent [4] place!

The sheriffs of the West, 1883.
(U.S. National Archives.)

1. **reward** : （此處指）幫助治安官捕獲罪犯所得到的酬金。
2. **terrorized** : 使驚恐不安。
3. **hanged** :
4. **violent** : 充滿暴力的。

KET

Read the paragraph. Choose the best word (A, B or C) for each space.

There were [0]..... outlaws in the West. [1]..... robbed banks and trains. There were [2]..... sheriffs.
Wyatt and Virgil Earp were the heroes of the gunfight [3]..... the O.K. Corral. This gunfight [4]..... a legend. Billy the Kid [5]..... a killer and a cattle thief. Jesse James and [6]..... brother Frank were bank and train robbers. [7]..... the West outlaws were [8]..... hanged.

	A	B	C
0.	**A** lots	(**B**) a lot of	**C** much
1.	**A** Their	**B** Them	**C** They
2.	**A** any	**B** few	**C** none
3.	**A** at	**B** of	**C** for
4.	**A** become	**B** became	**C** becoming
5.	**A** be	**B** were	**C** was
6.	**A** his	**B** him	**C** he
7.	**A** By	**B** For	**C** In
8.	**A** usually	**B** always	**C** never

LOOKING AT PICTURES
Look at the Reward Poster on page 47 and answer these questions.

a. How old was Billy the Kid?

....

b. How tall was he?

....

c. What color were his eyes?

....

d. Who was the sheriff?

....

e. How much was the reward for the capture of Billy the Kid?

....

3 **Identify the outlaws!**
How much do you know about the outlaws of the West? Complete their identities.

Outlaws		
Name	.... the	.... James
Description	Killer and	Bank and ....
Reward	$	$ ....
Killed by	....	....
Year	....	....
Place	....	....

4 **How many words with three or more letters can you make with this title?**

THE AMERICAN WEST

Two are done for you.

tea........	....	....	....	....
can........	....	....	....	....
....	....	....	....	....
....	....	....	....	....

The California Gold Rush

In 1848 John Marshall discovered gold at Sutter's Fort in California. This western territory became famous. Thousands of people traveled to California to look for gold. This was the California Gold Rush. These people were called "Forty-Niners".

The first "Forty-Niners" arrived in San Francisco in February 1849 on the steamship *California*. Others came to California on the Oregon and California Trail. In 1849 almost 100,000 people arrived in California to look for gold! San Francisco became a very important city.

END

There were many small mining[1] towns in the California Gold Country. Many "Forty-Niners" found gold, but only some of them became rich and important.

1. **mining**：淘金的。

The discovery of gold changed the destiny[1] of California and the West. The population of California grew very quickly. It became the 31st state of the United States in 1850.

A lot of people traveled to the West. A lot of gold traveled from California to the East. From 1852 the Wells Fargo stagecoaches traveled from the East Coast to the West Coast. Wells Fargo was a very important company in the West. The stagecoaches carried passengers,[2] money, mail and gold across the continent.

A journey across America by stagecoach was a great adventure!

Four or six horses pulled the Wells Fargo stagecoaches. There were four to six passengers in every stagecoach. The journey was very uncomfortable. The stagecoaches traveled all day and all night. Passengers slept inside the coaches on hard seats. Indians and outlaws attacked the stagecoaches.

A Wells Fargo office in San Francisco, 1860s.

1. **destiny** : 命運。
2. **passengers** : 旅客。

UNDERSTANDING THE TEXT

KET

Are these sentences "Right" (A) or "Wrong" (B)? If there is not enough information to answer "Right" (A) or "Wrong" (B), choose "Doesn't say" (C).

0. John Marshall discovered gold in San Francisco in 1848.

(**A**) Right **B** Wrong **C** Doesn't say

1. The "Forty-Niners" arrived in San Francisco by train.

A Right **B** Wrong **C** Doesn't say

2. During the Gold Rush the population of California grew quickly.

A Right **B** Wrong **C** Doesn't say

3. The biggest mining town was Placerville.

A Right **B** Wrong **C** Doesn't say

4. California became an American state in 1850.

A Right **B** Wrong **C** Doesn't say

5. Wells Fargo was an important stagecoach company.

A Right **B** Wrong **C** Doesn't say

6. Wells Fargo had two hundred stagecoaches.

A Right **B** Wrong **C** Doesn't say

7. Stagecoaches traveled during the day and stopped at night.

A Right **B** Wrong **C** Doesn't say

Unscramble these sentences.

a. California / were / many / in / towns / mining / there.

b. gold / "Forty-Niners" / found / many.

c. rich / became / important / and / some.

d. carried / money / stagecoaches / passengers / and.

e. was / the / uncomfortable / journey / very.

f. the / attacked / Indians / stagecoaches / outlaws / and.

Have fun with this Western crossword puzzle!

ACROSS

1. In 1850 California became a ..... .
2. Gold was discovered here.
3. The "Forty-Niners" looked for ..... .
4. Important American city.
5. Important company in the West.

DOWN

6. They attacked stagecoaches.
7. The opposite of "out".
8. The California Gold ..... .
9. A famous trail.

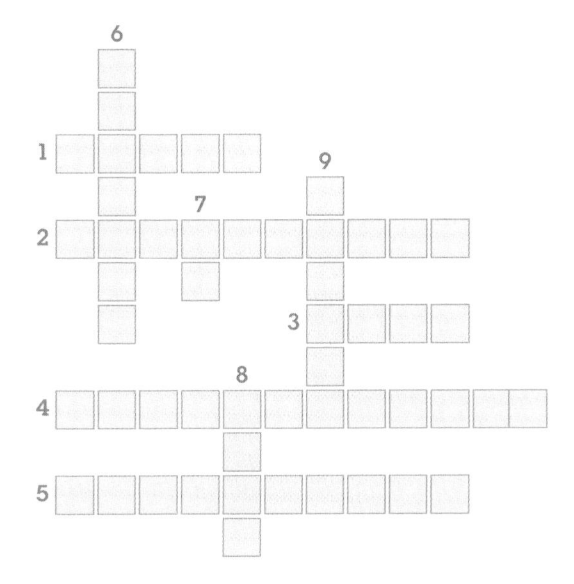

Listen to the first two paragraphs of Chapter 6. Then listen to them again and fill in the gaps.

In 1848 John Marshall discovered .... at Sutter's Fort in California. This western territory became famous. Thousands of people traveled to .... to look for gold. This was the California Gold Rush. These people were .... "Forty-Niners".

The first "Forty-Niners" .... in San Francisco in February 1849 on the steamship *California*. Others .... to California on the Oregon and California Trail. In 1849 almost 100,000 .... arrived in California to .... for gold! San Francisco became a very important ....

PROJECT ON THE WEB

Let's find out more about the Gold Rush!

Answer the following questions.

1 When and where was gold discovered?
2 Who discovered it?
3 Where is the Gold Country and what happened there?
4 What happened to the city of San Francisco when gold was discovered?
5 How did people get to the Gold Country?
6 What was life like in the Gold Country?

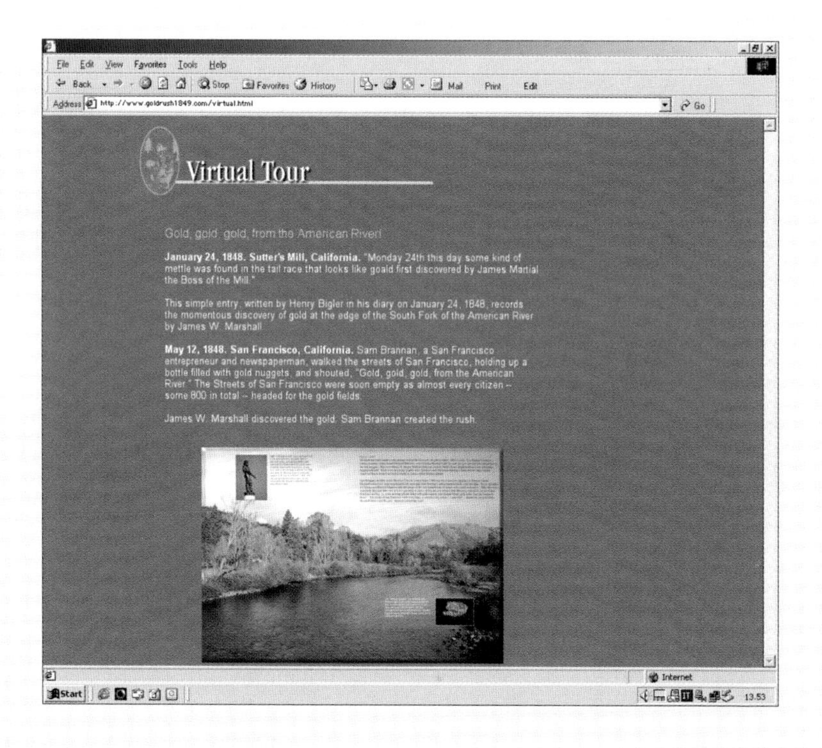

COWBOYS!

Everyone knows the word "cowboy"! The cowboy is an important part of American folklore.[1] There are many films about cowboys and the West.

The cowboys were young, adventurous[2] men. Many cowboys were white, but others were Mexicans or Afro-Americans. The cowboy's life was difficult. Cowboys followed a severe[3] code of conduct.[4] They lived in difficult conditions. They worked from morning to evening. Often they worked at night, too.

Cowboys worked on a ranch. They rode horses and looked after

Loops and Swift Horses are Surer than Lead by Charles M. Russell.
(Amon Carter Museum, Texas)

1. **folklore** : 民間傳說。
2. **adventurous** : 愛冒險的。
3. **severe** : 非常嚴格的。
4. **code of conduct** : 行為規範。

cattle, day and night. They moved the cattle from one grazing place[1] to another.

Cowboys took the cattle to the nearest railroad station. This was often about 1,000 miles (1,600 kilometers) away! It took many weeks to arrive. At the railroad station the cowboys sold the cattle. The cattle went by train to the East. Americans liked good meat from the West.

The cattle went by train to the East.

1. **grazing place** : 放牧地。

What did cowboys wear? A cowboy always wore a big hat. It protected him from the hot sun and the rain. He wore a bandana [1] around his neck. He used the bandana to protect his nose and mouth. He wore a shirt, a vest [2] and trousers. Every cowboy wore big boots and had a pistol. [3]

There are still cowboys in the United States today! They work on cattle ranches in the West and they ride horses.

Branding [4] *a yearling* [5] by Frank Sherman, early 1900s.

1. **bandana** : 
2. **vest** : 
3. **pistol** : 手槍。
4. **branding** : 給某物打烙印。
5. **yearling** : 一歲馬。

KET

1 **Read the descriptions.**
What is the word for each one?
The first letter is already there. There is one space for each other letter in the word.

0. It looks like a shirt but it has no sleeves. v e s t

1. You wear them on your feet. b _ _ _ _

2. You wear it on your head. h _ _

3. A cowboy put it on his mouth to protect his nose and mouth. b _ _ _ _ _ _

4. Every cowboy had one to defend himself. p _ _ _ _ _

5. You wear them on your legs. t _ _ _ _ _ _ _

2 **LOOKING AT PICTURES**
Look at the picture on page 57.

a. How many cowboys are there in the picture?

....

b. Where are they?

....

c. What are they doing?

....

d. What are they wearing?

....

The Cowboy by Frederic Remington, 1902.
(Amon Carter Museum, Texas)

The Indian Wars

The Indian Wars began in the 1800s. Thousands of pioneers went to the West. They took the Indians' land. The Indians were angry. They did not want to lose their land.

The Wars of the Midwest

In 1811 the Shawnee Indians of the Indiana Territory attacked the settlers and the U.S. Army.

These Indian Wars continued until 1832. The Black Hawk War was the last Indian war in the Midwest. The Indians lost this war and their land.

The Wars of the Southeast

In the 1830s Osceola was the chief of the Seminole Indians of Florida. He said, "We will fight until the last drop of Seminole blood!" The Seminole Indians did not want to leave their land. They fought for many years. In 1837 Chief Osceola was captured. Most of the Seminoles were killed.

The Cherokee Indians were a strong and important tribe. They had big plantations [1] and farms. The settlers wanted their land. In 1839 the U.S. Government told the Cherokees to leave their homes. They went to an Indian reservation in Oklahoma. Many Indians died during the long journey. The journey was called the Trail of Tears. [2]

The Wars of the Great Plains

In the 1850s many pioneers crossed the Great Plains. Many of them settled there. The Sioux and Cheyenne Indians fought against these settlers.

The Sioux and Cheyenne Indians were courageous warriors. [3] The Sioux Chiefs were Crazy Horse, Red Cloud and Sitting Bull. They told their warriors, "Fight to kill, or you will lose your lands!"

The Sioux and Cheyenne Indians attacked U.S. Army forts and settlers. The fighting was terrible. It continued for many years.

The U.S. Government told the Indians to go to an Indian reservation. Chiefs Crazy Horse and Sitting Bull did not want to go. There were a lot of brutal [4] battles. Many people were killed.

The Apache and Comanche Indians also fought against the U.S. Army and the settlers for many years.

The Apaches were great warriors. They loved their land and their freedom. They didn't want to lose them. They preferred to fight and die. Geronimo and Cochise were famous Apache chiefs. Their courage was legendary. [5] Everyone was afraid of Geronimo and Cochise. The Indian Wars ended in 1890.

1. **plantations** : 種植園。
2. **tears** :
3. **warriors** : 戰士。
4. **brutal** : 殘酷的。
5. **legendary** : 傳為佳話的。

The Massacre[1] *of Sand Creek, Colorado Territory, 1864,* by Robert Lindneux, 1936. (Colorado Historical Society)

1. **massacre**：大屠殺。

UNDERSTANDING THE TEXT

KET

Read the paragraph. Choose the best word (A, B or C) for each space.

The pioneers [0].A.. the Indians' land [1]..... the West. The Indians [2]..... angry and the Indian Wars [3]..... .

The Indian Wars in the Midwest continued [4]..... 1832. The Cherokee Indians left [5]..... homes in 1839 and went to a reservation in Oklahoma. They were an important [6]..... . The Sioux, Cheyenne, Apache and Comanche Indians [7]..... for many years. [8]..... 1890 the Indian Wars ended.

0.	(**A**) took	**B** taking	**C** takes
1.	**A** into	**B** in	**C** on
2.	**A** is	**B** was	**C** were
3.	**A** began	**B** beginning	**C** begins
4.	**A** in	**B** until	**C** for
5.	**A** their	**B** theirs	**C** its
6.	**A** family	**B** tribe	**C** group
7.	**A** fighting	**B** fight	**C** fought
8.	**A** On	**B** In	**C** At

LOOKING AT PICTURES

Look at the painting on pages 62-63 and answer the following questions.

a. What's happening in the painting?

b. Who's riding the horses?

c. Who's on foot?

d. What color is the uniform of the American soldiers?

e. Who's winning the battle? Why?

3 **Look at the map of the United States and fill in the names of the tribes in the correct places.**

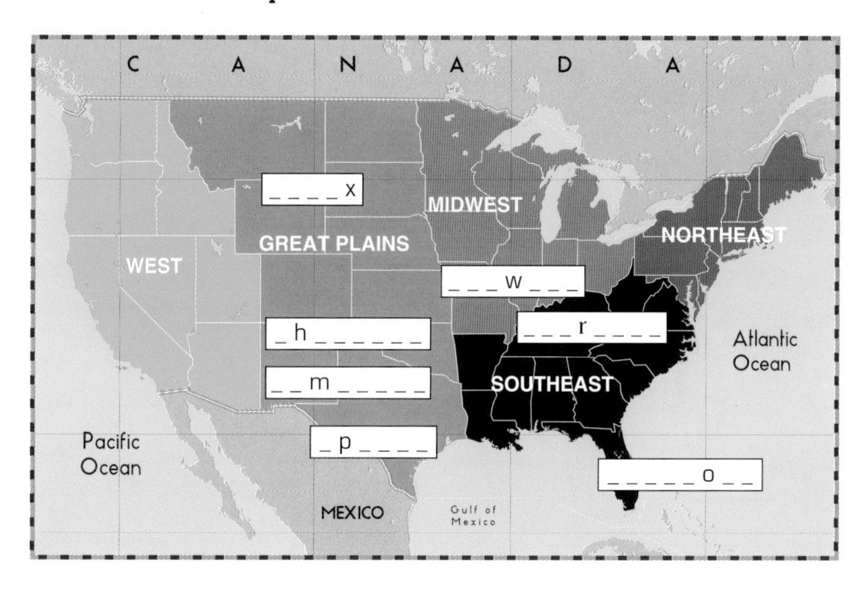

4 **Let's make a short summary of Chapter 7. Fill in the gaps with the correct words from the box. Some words can be used more than once.**

land   Great Plains   last   Indian   chiefs
reservation   ended   forts   tribes

The .... Wars began in the 1800s. Thousands of pioneers went to the West and took the Indians' .... There were many wars. The Black Hawk War was the .... Indian war in the Midwest.

The Seminole and Cherokee were important .... They lost their .... and went to live on a .... in Oklahoma.

The Sioux and Cheyenne Indians lived on the .... They attacked U.S. Army .... and settlers.

Geronimo and Cochise were great Apache .... The Indian Wars .... in 1890.

EXIT TEST

1 WHO ARE THEY?
Match the numbers with the letters.

WHO...

1. ☐ opened the road to the West?
2. ☐ bought the Louisiana territory?
3. ☐ explored the Louisiana Purchase?
4. ☐ was a friendly Indian woman?
5. ☐ was a famous sheriff?
6. ☐ were the heroes of the O.K. Corral?
7. ☐ was a dangerous outlaw?
8. ☐ discovered gold at Sutter's Fort?
9. ☐ looked for gold in California?
10. ☐ rode horses and looked after cattle?
11. ☐ was Sitting Bull?
12. ☐ went to live on government reservations?
13. ☐ was a famous Apache chief?

A a Sioux chief
B John Marshall
C Wyatt and Virgil Earp
D the American Indians
E Geronimo
F Daniel Boone
G President Thomas Jefferson
H Pat Garrett
I Billy the Kid
J cowboys
K the "Forty-Niners"
L Sacagawea
M Lewis and Clark

KET

2 **Are these sentences "Right" (A) or "Wrong" (B)? If there is not enough information to answer "Right" (A) or "Wrong" (B), choose "Doesn't say" (C).**

0. The American Indians came from South America.
A Right (**B**) Wrong **C** Doesn't say

1. All American Indians were hunters.
A Right **B** Wrong **C** Doesn't say

2. Daniel Boone was a prisoner of the Shawnee Indians.
A Right **B** Wrong **C** Doesn't say

3. Boone's wife was a school teacher.
A Right **B** Wrong **C** Doesn't say

4. The population of America grew very quickly.
A Right **B** Wrong **C** Doesn't say

5. President Jefferson was interested in science.
A Right **B** Wrong **C** Doesn't say

6. The Lewis and Clark expedition explored the Southeast.
A Right **B** Wrong **C** Doesn't say

7. It was a very successful expedition.
A Right **B** Wrong **C** Doesn't say

8. The Oregon and Santa Fe Trails took the pioneers to the West.
A Right **B** Wrong **C** Doesn't say

9. Fort Laramie was the biggest fort in the West.
A Right **B** Wrong **C** Doesn't say

10. All "Forty-Niners" found gold and became rich.
A Right **B** Wrong **C** Doesn't say

11. The last Indian War in the Midwest was The Black Hawk War.
A Right **B** Wrong **C** Doesn't say

12. The Sioux and Cheyenne Indians lived in the Great Plains.
A Right **B** Wrong **C** Doesn't say

13. The Apache and Comanche Indians did not fight in the Indian Wars.
A Right **B** Wrong **C** Doesn't say

Which notice (A-H) says this (1-5)?

1. You can buy a vest and a bandana here.
2. Be careful of Indian attacks.
3. You can go to the bank on Thursday.
4. You mustn't cross the river.
5. You can sell your gold here.

1. ☐ 2. ☐ 3. ☐ 4. ☐ 5. ☐

4 Fill in the gaps with the words from the box.

million four hundred wilderness fifty grew
Missouri independent Mississippi Indians

1. The United States of America has .... states.
2. About 250 .... people live in the United States.
3. The .... and the .... are two important American rivers.
4. The first European explorers came to America about .... years ago.
5. At that time America was an enormous .... and .... lived there.
6. America became .... from Britain in 1776 and .... very quickly.

5 Are the following sentences true (T) or false (F)? Correct the false ones.

	T	F
a. The American Indians came from Asia about 40,000 years ago.	☐	☐
b. The Indians were farmers and explorers.	☐	☐
c. The European settlers and explorers took the land from the Indians.	☐	☐
d. Daniel Boone opened the road to the South.	☐	☐
e. In 1803 President Thomas Jefferson bought the Louisiana territory from Britain.	☐	☐
f. Lewis and Clark explored the Louisiana Purchase and made maps. They were the first Indians to see this land.	☐	☐
g. Thousands of pioneers went to the East on the Oregon Trail and the Santa Fe Trail. The journey was short.	☐	☐
h. In the West there were a lot of outlaws but few sheriffs.	☐	☐
i. Billy the Kid and Jesse James were two famous sheriffs.	☐	☐
j. After the discovery of gold at Sutter's Fort in 1848 California became an important American state.	☐	☐

k. Wells Fargo stagecoaches carried passengers, money, gold and mail across the American continent. ☐ ☐

l. Cowboys worked in a town and moved cattle from one place to another. ☐ ☐

m. During the 1800's there were many Indian Wars because the Indians did not want to lose their land. ☐ ☐

n. The Apache, Comanche, Cheyenne and Sioux Indians were courageous warriors and lived in the Great Plains. ☐ ☐

6 Put the verbs into the Past Simple tense.

1. About twenty million Indians (live) .... in America.
2. The Indians (teach) .... the explorers how to grow corn, potatoes and tobacco.
3. Sacagawea (speak) .... English and other Indian languages.
4. Davy Crockett (fight) .... for the independence of Texas.
5. The pioneers (begin) .... their journey in spring.
6. The settlers (build) .... log cabins.
7. During the California Gold Rush a lot of people (go) .... to California.
8. The cowboys (sell) .... the cattle at the railroad station.

7 Fill in the gaps with the prepositions in the box.

on inside by in at after between

1. .... the Lewis Clark expedition many people went to the West.
2. The Lewis and Clark expedition traveled 12,800 kilometers .... 1804 and 1806.
3. The pioneers traveled .... covered wagons.
4. They traveled .... the Oregon Trail.
5. The covered wagons were pulled .... mules or oxen.
6. Pioneers got up .... 5 a.m. and began to travel.
7. Passengers slept .... the stagecoaches on uncomfortable seats.

8 **Tick the correct answers.**

1.

☐ "Forty-Niners"

☐ Cowboys

looked for gold in California.

2. The "circuit rider" was

☐ a religious man.

☐ an outlaw.

3. The leader of a wagon train was called

☐ a scout.

☐ a captain.

4. The population of America grew

☐ slowly.

☐ quickly.

5. The first Americans were

☐ the Indians.

☐ the explorers.

9 **Which part of The American West is the most interesting?**

....

....

10 **What part of America do you want to visit?**

....

....

11 **Do you like films about cowboys and the West?**

....

....

KEY TO THE EXERCISES AND EXIT TEST

KEY TO THE EXERCISES

A NATION IS BORN

Page 9 Exercise 1
1. B **2.** B **3.** A **4.** C **5.** A **6.** C

Page 11 Exercise 2
1. B **2.** A **3.** C **4.** B **5.** A **6.** C **7.** C

CHAPTER ONE

Page 16 Exercise 1
1. C **2.** C **3.** B **4.** A **5.** C **6.** A **7.** B
8. A **9.** A

Page 17 Exercise 2
lose – lost
take – took
want – wanted
love – loved
use – used
follow – followed
ride – rode
die – died
bring – brought
teach – taught
be – was/were
have – had
live – lived
go – went
come – came

Page 17 Exercise 3
a. lived **b.** had **c.** taught **d.**brought
e. rode **f.** wanted **g.** lost

Page 18 Exercise 4
a. reservations **b.** hunters **c.** food
d. canoe **e.** chief **f.** tools

Page 18 Exercise 5
a. tools **b.** chief **c.** hunters **d.** food
e. canoe **f.** reservations

Page 18 Exercise 6
Open questions.

CHAPTER TWO

Page 21 Exercise 1
1. C **2.** C **3.** A **4.** C **5.** A

Page 21 Exercise 2
a. Daniel Boone and the settlers.
b. Through the Cumberland Gap.
c. Autumn.
d. It's cloudy.

Page 22 Exercise 3
1. A **2.** A **3.** B **4.** A **5.** C

Page 23 Exercise 4
road, fort, loved, free, went, many, West

Page 23 Exercise 5

M F P X C U O H Y D U R J F
M I S S O U R I I D A Q L G
B E G U T J D C X J H L Q S
G A P P A L A C H I A N R C
J T K E F G E A P K G X T D
E G T R T F X F R E A P S S
A V M I S S I S S I P P I H
X F R O A R H C X N M O A D
O H A R O C K Y M O A D N K
Y W S Q O C T O E C J W A Z

CHAPTER THREE

Page 27 Exercise 1
a. bought **b.** explore **c.** fifty, up
d. winter **e.** woman **f.** Ocean;
g. explorers

Page 27 Exercise 2
a. 5 **b.** 4 **c.** 6 **d.** 1 **e.** 3 **f.** 2.

Page 28 Exercise 3
M. Lewis and W. Clark: courageous, explorers, made maps, built a fort
Sacagawea: friendly, helpful, Indian, spoke many Indian languages

CHAPTER FOUR

Page 34 Exercise 1
1. A **2.** B **3.** C **4.** A **5.** A **6.** B **7.** A
8. C **9.** B

Page 34 Exercise 2
Open questions.

Page 35 Exercise 3
1. E **2.** H **3.** G **4.** A **5.** D

Page 35 Exercise 4
a. scientist **b.** poor **c.** trail
d. wagon **e.** fun

Page 36 Exercise 5
1. settler **2.** trapper **3.** captain
4. scout **5.** mule

Page 36 Exercise 6
Open questions.

Page 37 Exercise 1
1. C **2.** B **3.** A **4.** C **5.** B **6.** C
7. A

CHAPTER FIVE

Page 42 Exercise 1
1. B **2.** B **3.** A **4.** C **5.** B **6.** A

Page 43 Exercises 2-3
Open questions.

Page 44 Exercise 4
1. on **2.** for **3.** by **4.** was
5. There **6.** My **7.** all **8.** to **9.** in
10. were

Page 44 Exercise 5
Open questions.

Page 45 Exercise 6
a. 3 **b.** 5 **c.** 6 **d.** 1 **e.** 2 **f.** 4.

OUTLAWS OF THE WEST

Page 48 Exercise 1
1. C **2.** B **3.** A **4.** B **5.** C **6.** A **7.** C
8. A

Page 48 Exercise 2
a. 18 **b.** 5 feet 3 inches **c.** Blue
d. Jim Dalton **e.** $5,000.00

Page 49 Exercise 3

Name	Billy the Kid	Jesse James
Description	Killer and cattle thief	Bank and train robber
Reward	$ 5,000.00	$ 10,000.00
Killed by	Pat Garrett	Bob Ford
Year	1881	1882
Place	New Mexico	Missouri

Page 49 Exercise 4
Open question.

CHAPTER SIX

Page 52 Exercise 1
1. B **2.** A **3.** C **4.** A **5.** A **6.** C **7.** B

Page 52 Exercise 1
a. There were many mining towns in California.
b. Many "Forty-Niners" found gold.
c. Some became rich and important.
d. Stagecoaches carried money and passengers.
e. The journey was very uncomfortable.
f. Indians and outlaws attacked the stagecoaches.

Page 53 Exercise 3

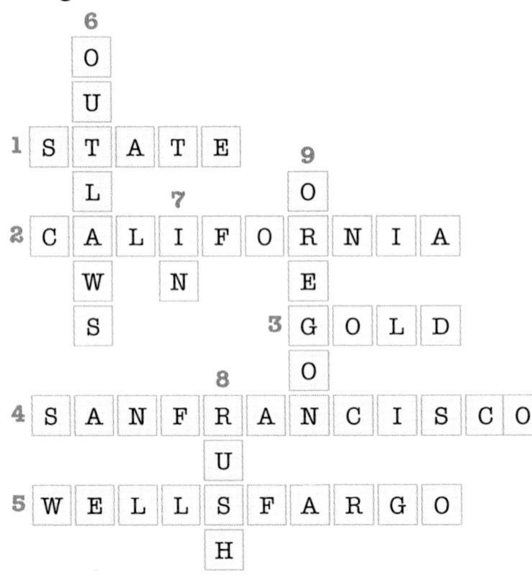

Page 53 Exercise 4
gold, California, called, arrived, came, people, look, city

COWBOYS!

Page 58 Exercise 1
1. boots **2.** hat **3.** bandana **4.** pistol **5.** trousers

Page 58 Exercise 2
a. Five.
b. In a field.
c. They are "branding a yearling".
d. Cowboy hats, vests, shirts and boots. They have bandanas around their necks.

CHAPTER SEVEN

Page 64 Exercise 1
1. B **2.** C **3.** A **4.** B **5.** A **6.** B **7.** C **8.** B

Page 64 Exercise 2
Open questions.

Page 65 Exercise 3

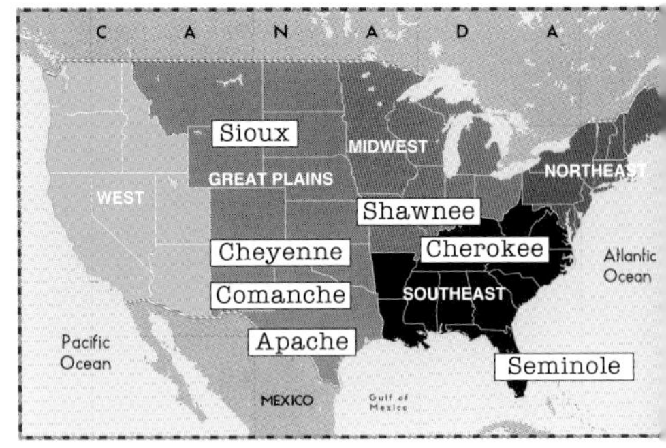

Page 65 Exercise 4
Indian, land, last, tribes, land, reservation, Great Plains, forts, chiefs, ended

KEY TO THE EXIT TEST

1. F **2.** G **3.** M **4.** L **5.** H **6.** C **7.** I
8. B **9.** K **10.** J **11.** A **12.** D **13.** E

1. B **2.** A **3.** C **4.** A **5.** A **6.** B **7.** A
8. A **9.** C **10.** B **11.** A **12.** A **13.** B

1. G **2.** C **3.** D **4.** A **5.** F

1. fifty
2. million
3. Missouri / Mississippi
4. four hundred
5. wilderness / Indians
6. independent / grew

a. True.
b. False. The Indians were farmers and hunters.
c. True.
d. False. Daniel Boone opened the road to the West.
e. False. In 1803, President Thomas Jefferson bought the Louisiana territory from France.
f. False. Lewis and Clark explored the Louisiana Purchase and made maps. They were the first white men to see this land.
g. False. Thousands of pioneers went to the West on the Oregon Trail and the Santa Fe Trail. The journey was long.
h. True.
i. False. Billy the Kid and Jesse James were two famous outlaws.
j. True.
k. True.
l. False. Cowboys worked on a ranch and moved cattle from one place to another.
m. True.
n. True.

1. lived
2. taught
3. spoke
4. fought
5. began
6. built
7. went
8. sold

1. After
2. between
3. in
4. on
5. by
6. at
7. inside

1. "Forty-Niners"
2. a religious man
3. a scout
4. quickly
5. the Indians

Open questions.

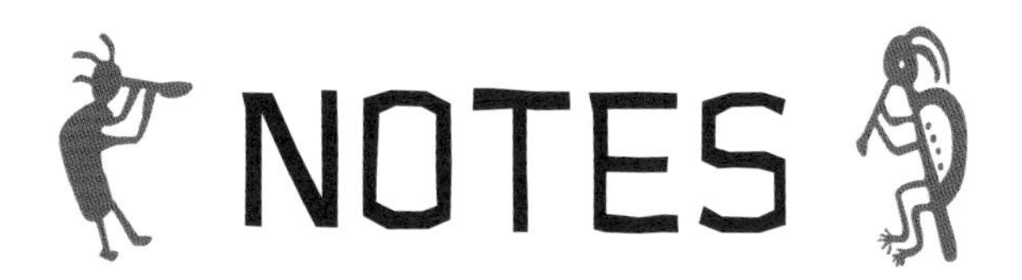

NOTES

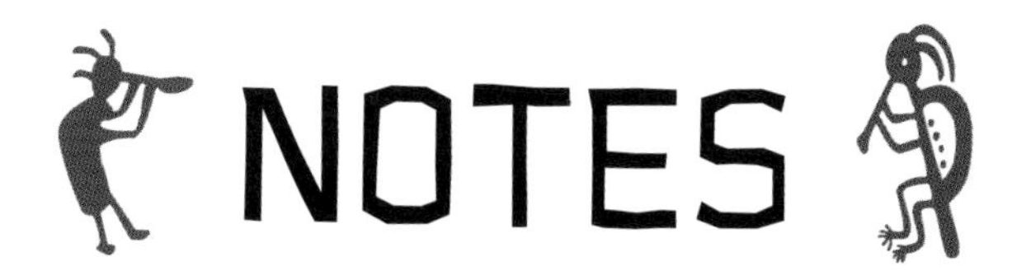
NOTES

This Chinese edition of *The American West*
has been published with the written permission of
Black Cat Publishing.

Name of Book: The American West
Author: Gina D.B. Clemen
Editor: Monika Marszewska
Design and art direction: Nadia Maestri
Computer graphics: Simona Corniola
Illustrations: Franco Grazioli
Edition: © 2003 Black Cat Publishing
an imprint of Cideb Editrice, Genoa, Canterbury

系 列 名：Black Cat 優質英語階梯閱讀 · Level 1
書　　名：美國西部探險
責任編輯：傅　伊
封面設計：張　毅
出　　版：商務印書館（香港）有限公司
香港筲箕灣耀興道 3 號東滙廣場 8 樓
http://www.commercialpress.com.hk
印　　刷：中華商務彩色印刷有限公司
香港新界大埔汀麗路 36 號中華商務印刷大廈
版　　次：2004 年 2 月第 1 版第 1 次印刷

ISBN 962 07 1691 4
Printed in Hong Kong

BLACK CAT ENGLISH CLUB

Membership Application Form

BLACK CAT ENGLISH CLUB is for those who love English reading and seek for better English to share and learn with fun together.

Benefits offered:
- *Membership Card*
- *Book discount coupon*
- *English learning e-forum*
- *English learning activities*
- *Black Cat English Reward Scheme*
- *Surprise gift and more...*

Simply fill out the application form below and fax it back to 2565 1113 or send it back to the address at the back.

Join Now! It's FREE exclusively for readers who have purchased *Black Cat English Readers* **!**

(Please fill out the form with **BLOCK LETTERS**.)

The title of Black Cat English Reader/book set that you have purchased: ____________________

English Name: ________________ (Surname) ________________________ (Given Name)

Chinese Name: ____

Address: ____

____

____

Tel: ________________________________ Fax: ________________________________

Email: ____

(Login password for e-forum will be sent to this email address.)

Sex: ❑ Male ❑ Female

Education Background: ❑ Primary 1-3 ❑ Primary 4-6 ❑ Junior Secondary Education (F1-3)
❑ Senior Secondary Education (F4-5) ❑ Matriculation
❑ College ❑ University or above

Age: ❑ 6 - 9 ❑ 10 - 12 ❑ 13 - 15 ❑ 16 - 18 ❑ 19 - 24 ❑ 25 - 34
❑ 35 - 44 ❑ 45 - 54 ❑ 55 or above

Occupation: ❑ Student ❑ Teacher ❑ White Collar ❑ Blue Collar
❑ Professional ❑ Manager ❑ Business Owner ❑ Housewife
❑ Others (please specify: ________________)

As a member, what would you like **BLACK CAT ENGLISH CLUB** to offer:

❑ Member gathering/ party ❑ English class with native teacher ❑ English competition
❑ Newsletter ❑ Online sharing ❑ Book fair
❑ Book discount ❑ Others (please specify: ________________________)

Other suggestions to **BLACK CAT ENGLISH CLUB**: ________________________________

____

Please sign here: ________________________ (Date: ________________________)

Visit us at Quality English Learning Online http://publish.commercialpress.com.hk/qel

Stamp
Here

BLACK CAT ENGLISH CLUB
The Commercial Press (Hong Kong) Ltd.
9/F, Eastern Central Plaza,
3 Yiu Hing Road, Shau Kei Wan,
Hong Kong